Water-Saving Tips

DUMMIES®

Water-Saving Tips FOR DUMMIES®

by Michael Grosvenor

WILEY

Wiley Publishing Australia Pty Ltd

Water-Saving Tips For Dummies®

Published by
Wiley Publishing Australia Pty Ltd
42 McDougall Street
Milton, Qld 4064
www.dummies.com

National Library of Australia
Cataloguing-in-Publication data

Grosvenor, Michael, 1966–.
Water-saving tips for dummies.

Includes index.
ISBN 978 0 731 40775 0 (pbk.).

1. Water efficiency. 2. Water conservation. 3. Household ecology.
4. Water reuse. I. Title. II. Title: Water-saving tips for dummies.
(Series: For dummies).

333.9116

Printed in China by
Printplus Limited

10 9 8 7 6 5 4 3 2 1

About the Author

Michael Grosvenor is a leading urban planning professional and freelance writer on sustainability. Through his work and writing, Michael promotes the benefits of making sustainable lifestyle choices. Michael has particular expertise in transport and advises the private sector and government on policies that promote increased public transport, walking and cycling facilities. Michael is a strong advocate for the important role that public transport plays in our cities and towns. He is the author of *Sustainable Living For Dummies* and *Energy-Saving Tips For Dummies*.

Michael is the director of his own consultancy and holds Masters degrees in Urban Affairs and Applied Social Research and a Degree in Town Planning. He is also a member of the Planning Institute of Australia and provides advice to the Institute on integrated land use and transport planning issues.

Michael has lived and studied in New York City, and he currently enjoys an inner-city lifestyle in Sydney, Australia.

Dedication

To my best friend and partner, Justine — thank you for your encouragement and support.

Author's Acknowledgments

My desire to talk to the general public about sustainable living motivated me to write this book. I'm often preaching to the converted in my consulting work. The environmentalists, planners, architects, social scientists, engineers and geographers I work with find ourselves saying the same things to each other, and we're often scribbling messages and ideas on whiteboards that no-one else gets to see.

Writing a book for an audience interested in adopting water-saving tips in the home has been very rewarding. This book covers a lot of useful information about water-wise ways — perhaps too much ground for one person to have the required expertise on every topic. I have been able to carry out the necessary research for this book thanks to the thousands of committed professionals out there who have tested, researched and published their findings about the problems facing the planet. This book could not be written without their passion.

I thoroughly enjoyed working with editors Robi van Nooten and Maryanne Phillips and thank them both for excellent editorial contributions and ideas.

Publisher's Acknowledgments

We're proud of this book; please send us your comments through our online registration form located at www.dummies.com/register/.

Some of the people who helped bring this book to market include the following:

Acquisitions, Editorial and Media Development

Project Editors: Robi van Nooten, On-Track Editorial Services; Maryanne Phillips

Acquisitions Editor: Charlotte Duff

Editorial Manager: Gabrielle Packman

Image credits:

- BlueScope Steel Limited: page 16, © BlueScope Water
- Commonwealth Copyright Administration: page 28, Department of Environment & Heritage Water Efficiency Labelling and Standards (WELS) Scheme. © Commonwealth of Australia, reproduced by permission
- Solahart Industries Pty Ltd: page 31

Production

Layout and Graphics: Wiley Composition Services, Wiley Art Studio

Cartoons: Glenn Lumsden

Proofreader: Marguerite Thomas

Indexer: Max McMaster, Master Indexing

Contents at a Glance

Introduction *1*

Chapter 1: Water: A Precious Resource 7
Chapter 2: Saving Water in the Home 25
Chapter 3: Greening Up Your Garden the Water-Wise Way 33
Chapter 4: Ten Water-Saving Tips 53

Index *59*

Table of Contents

Introduction 1

How to Use This Book 2
How This Book Is Organised 2
Chapter 1: Water: A Precious Resource 2
Chapter 2: Saving Water in the Home 3
Chapter 3: Greening Up Your Garden the Water-Wise Way 3
Chapter 4: Ten Water-Saving Tips 3
Icons Used in This Book 4

Chapter 1: Water: A Precious Resource 7

Understanding Why Water Is a Precious Commodity 8
Where your water comes from 8
Where your wastewater goes 9
Adopting a Drier Lifestyle 11
Eliminating wastage 12
Tanking it 15
How grey is your water? 16
Drinking the Stuff 18
How safe is tap water? 18
Bottled water is 90 per cent energy 19
Australia, You're Baking In It — Adapting to a Drier Climate 20
Recycling urban wastewater 21
Removing the salt 22
Coping with Agricultural Water Shortages 22
Impact on agriculture 23
Impact on remote communities 24

Chapter 2: Saving Water in the Home 25

Bathroom Blitz 25
Kitchen Connection 28
Laundry Time: Doing an Eco-Wash 29
Using the Sun to Heat Your Water 30

Chapter 3: Greening Up Your Garden the Water-Wise Way 33

Designing Your Water-Wise Garden 34
Planning the landscape 35
Planning what to plant 36
Working with the lay of the land 37
Maintaining a Sustainable Garden 38
Minding your water usage 38
Keeping a check on your garden's health 39
Improving the chemistry for a healthier soil 40
Growing food naturally 42
Composting Your Waste 43
Beginning an outdoor compost 45
Setting up a worm farm 46
Choosing and Planting Native Shrubs and Plants 48
Living With a Lawn 50

Chapter 4: Ten Water-Saving Tips 53

Turn It Off 53
Check the Plumbing 54
Switch to Water-Saving Taps and Shower Heads 54
Look at the Ratings 54
Catch It Before It Goes Down the Drain 55
Fixing Your Toilet Habits 55
Save Every Last Drop 56
Recycle Your Greywater 56
Make Better Use of the Lawn 56
Garden Naturally 57

***Index* 59**

Introduction

Most people agree that humans are using natural resources faster than the environment is able to regenerate them. Water is one of the most precious natural resources, but by adopting some simple lifestyle choices and habits that include making better use of water, you can help to get the planet's balance back on an even keel.

More than likely you picked up this book because you know the limitations of such a valuable natural resource as water. *Water-Saving Tips For Dummies* provides you with practical tips that you need to conserve, reuse and recycle what goes down your drain. This book deals with water issues on the grand scale of living in a drier country as well as on the home scale of the bathroom, kitchen, laundry and garden. This book isn't about changing your whole life by denying a certain lifestyle; instead, it offers practical advice about how to live in a more water-wise fashion.

I'm not expecting you to adopt every single one of the actions contained in this book. Far from it. I know from personal experience that doing so is a life-long task. Starting small by adopting any one of the tips in this book is a positive beginning.

Significantly, this book isn't just about the actions you take in the bathroom, kitchen, garden and laundry. This book may be the beginning of a journey that leads you to make other choices as a consumer about embracing sustainability in general. I'm confident that this book can give you a better understanding of the choices available to you and how making these choices can make a real difference.

How to Use This Book

I'd love you to sit down and read the book from beginning to end. After all, I think it's a great read. But realistically, you're probably likely to dip in and out of it to find the tips you need. That's why *Water-Saving Tips For Dummies* covers each topic in its own chapter. You can skim the contents and go straight to the chapters that interest you most.

If you're coming to this book as a water-saving novice and want a bit of background on the topic, your best bet is to go straight to Chapter 1. You can then move through subsequent chapters (and rooms of the house — through to the garden) at your own speed with a much better feel for why each action is worthwhile.

How This Book Is Organised

Water-Saving Tips For Dummies is divided into four chapters, each focusing on a different aspect of how to make better use of water in your daily life.

Chapter 1: Water: A Precious Resource

This chapter describes why water is a precious natural resource, globally and at home. I lay the foundation for how we get access to water and how we drain it away, and why these methods have to change simply because not enough water is available in our dry-climate country to supply increasing demands.

Chapter 2: Saving Water in the Home

The three *R*s — reducing, reusing and recycling — play a critical role in a water-wise household. This chapter contains tips that help lower the amount of waste in the bathroom, kitchen and laundry by encouraging you to adopt lifestyle changes that make a difference.

Chapter 3: Greening Up Your Garden the Water-Wise Way

Your garden has potential plus when it comes to making a positive influence on the environment. In this chapter I show you how to reduce your reliance on water by planning the landscape and choosing appropriate plants. A sustainable garden reduces water wastage.

Chapter 4: Ten Water-Saving Tips

This helpful chapter gives you my top-ten tips for saving water. And, if your kids read just this chapter, they're likely to grow up with a water-wise mindset.

Icons Used in This Book

This icon highlights inside stories about real people who incorporate water-saving practices into their lifestyle.

This icon flags handy Web sites.

Warning icons are serious stuff. Read carefully and take heed.

Don't forget these little pearls of wisdom. Remember and remember . . .

This icon flags relatively in-depth detail. You may want to skim over these, or undertake further research on the technical area being discussed. Believe me, they can be fascinating.

Tips are the little things you can do to make your water-wise lifestyle more achievable. These brainwaves offer you handy shortcuts.

'Honey, it's the plants from next-door. Can they come in for a drink of water?'

Chapter 1

Water: A Precious Resource

In This Chapter

- Understanding how water is supplied to your home
- Using water wisely
- Drinking tap water versus bottled water
- Living with drought
- Coping with reduced rainfall in rural areas

Have you ever woken up and gone to the bathroom, turned on the tap, and discovered that nothing comes out? In your initial panic, you probably wondered how you were going to get through the next few hours — you can't take a shower, make breakfast or clean your teeth.

Your no-water crisis may have been caused by something like a broken pipe, which can be quickly fixed. Yet over one billion people in the world today have no guarantee of water from one day to the next. And if current trends continue, in less than 20 years, two-thirds of the world's population is going to suffer regular water shortages. Even the industrialised world is running out of water.

The bare facts are that simply not enough water is available to supply increasing demands. Since 1900, the world's population has grown two-fold, but global water use has increased six-fold. Reduced rainfall caused by climate change is now exacerbating the problem.

In this chapter, I explain how water supplies work and why the way water is delivered is going to change radically over the next ten years. I show you how these changes may affect you, and what you can do to make sure you're not left high and dry.

Understanding Why Water Is a Precious Commodity

Some facts about water can help you appreciate how dramatic the shortage situation is.

Up to 41 per cent of the world's population lives in regions that are under *water stress* — in these areas, water supply is much less than the global average. These parts of the world include northern and central Africa and the Middle East. By comparison, in urban areas of Australia, the United States and Europe, water supply is much higher than the global average. Despite this, the developed world also faces water shortages.

Where your water comes from

Ordinarily, water is delivered to homes and businesses in large urban areas as follows:

1. **Water from rivers and streams is collected in strategically placed dams and reservoirs, or drawn from natural underground sources.**

   Most big cities rely on many dams. Places such as Perth and Alice Springs get their water from underground sources.

2. **The water is delivered to filtration plants.**

   At the filtration plant, much of the sediment and minerals is removed, chlorine is added to kill any living things and fluoride is added to prevent tooth decay.

3. **The water is delivered to homes and businesses via a network of pumping stations and pipes.**

   These networks are extensive. For example, Sydney's network of underground water pipes is approximately 21,000 kilometres long — the same distance as flying from Sydney to New York via London.

For agricultural and farming purposes, the method is more direct: Irrigation water is pumped straight from a nearby watercourse or groundwater supply. Also, many rural towns pump their water supply straight from natural waterways and channels, and rely on residential rainwater tanks for back-up (for more details about rainwater tanks, see the section 'Tanking it' later in this chapter).

Many of the methods used to deliver water are unsustainable. Here's why:

- Damming natural waterways and concreting streams and rivers interferes with and, in some cases, destroys the natural ecosystem.
- Drawing too much water from rivers and streams reduces water flow; the removal of vegetation near rivers and streams intensifies sunlight; and agricultural run-off and sewage foul waterways with the wrong kind of nutrients. Algae blooms, which kill aquatic life, flourish under these conditions.
- Water drawn from rivers and underground water supplies for irrigation contains mineral salts. When this water evaporates, the salts are left in the top layers of the soil. This process increases *salinity*, which makes the soil useless for farming.
- Unless it rains, dams simply dry up. This drying up is accelerated by evaporation and by rates of rising consumption.
- Ageing pipes and infrastructures used to deliver water in urban areas are susceptible to leakages and contamination.

Where your wastewater goes

Getting water to your home, your business or your farm is one thing. But getting rid of the wastewater after you use it is all together another issue.

Who's drinking the dam dry?

According to research conducted by the CSIRO (Commonwealth Scientific and Industrial Research Organisation), Australia receives up to 3.3 million gigalitres of rainfall each year, but Australian homes and businesses consume only 20,000 gigalitres. This ratio suggests that plenty more water is available for everyone, right? Well, not really.

Of the 3.3 million gigalitres that Australia receives in rainfall each year, 88 per cent evaporates into the atmosphere.

Of the remaining 380,000 gigalitres, agriculture uses 80 per cent.

That leaves 72,000 gigalitres, but the majority of this disappears through leaky pipes. (In fact, in some urban areas, an incredible 80 per cent of the water supply is lost to leaks.)

The 20,000 gigalitres consumed in homes and industry is a large proportion of all the currently available water. Melbourne, Sydney, Brisbane, Perth and Adelaide all face serious water shortages in the next five years.

The term *wastewater* is used to describe all the water that has been used and then poured down the drain. Here are some of the more common components of wastewater:

- Water that runs down the kitchen sink
- Water that gets flushed down the toilet
- Water from showers and baths
- Water from washing machines
- Water used in industrial processes

This wastewater is also known as *sewage* and is distributed through sewerage pipes to plants near urban areas. These plants separate the gunk and sludge from the water. After treatment, the water is returned back into the environment via a nearby watercourse or the ocean. The separated sludge is usually disposed of in landfill or piped out into the ocean. Both processes have a negative impact on the environment.

Stormwater is the water that runs off roofs, streets and roads. Stormwater is managed separately to wastewater and usually runs straight into watercourses, along with the debris collected along the way, separated by grates and litter traps. The oil, grease and chemicals in stormwater cause up to half of the pollution in watercourses.

All this water can be recycled. Here's how:

- **Greywater:** You can safely reuse the wastewater from your washing machines, sinks and showers in your garden. This *greywater* contains relatively small amounts of pollutants and bacteria. (For information about reusing your greywater, see the section 'How grey is your water?' later in this chapter.)
- **Sewage:** Advances in sewage treatment systems allow the water separated at the sewage treatment plant to be put straight back into waterways or used in industry. (For more information about sewage recycling, see the section 'Recycling urban wastewater' later in this chapter.)
- **Stormwater run-off:** Although most stormwater currently runs straight into waterways or the sea, this source of water can potentially be stored and reused, with the gunk collected along the way separated from the water. At home, you can capture some of your own stormwater run-off from your roof by installing a rainwater tank. (For more details, see the section 'Tanking it' later in this chapter.)

Adopting a Drier Lifestyle

Governments around the world face dwindling water supplies and growing populations demanding more water. To resolve this issue, they're acting on two fronts:

- **Technological innovations:** Governments apply new technology to better manage the water supply. These developments enable water to be either recycled or drawn from new sources.
- **Behavioural measures:** Education and financial incentives encourage people to use water more conservatively to take the pressure off the water supply. In urban areas, 70 per cent of water consumption occurs in and around homes, so any changes in domestic consumption can have a significant impact.

Government actions aimed at changing your behaviour involve a combination of the 'carrot and the stick'. The *carrot* approach uses mainly financial incentives, designed to encourage you to conserve water. These include rebates for purchasing water-efficient washing machines and rainwater tanks. The *stick* approach forces you to do something, usually by introducing regulations that carry the threat of a fine. For example, water restrictions limit your use of water when supply levels get seriously low, and carry severe penalties. Watering gardens and washing cars are usually the first targets — these restrictions are permanent in some cities and towns. Another approach is raising the cost of water to discourage those who guzzle more than their fair share.

So how do you reduce your water consumption when temperatures rise? Actions you can take include the following:

- **Don't waste a drop:** Yes, and this resolve is all about being frugal with water. See the following section for more details.
- **Collect your rainwater:** Check out the section 'Tanking it' later in this chapter.
- **Recycle your own greywater:** For more details, see the section 'How grey is your water?' later in this chapter.

Eliminating wastage

Consuming less water isn't so difficult. By using water more efficiently, you can significantly cut your water consumption — and coincidentally your water bill.

Check for leaks regularly. If your toilet continues to run after flushing, or you have a dripping tap (inside or outside), call in the plumber. Just one leaking tap can waste up to 500 litres of water a week!

Place *aerators* on the faucets of all the taps in your home. This device can reduce water flow by up to 50 per cent without reducing water pressure. You can purchase a new water-saving tap aerator at most hardware stores. Just unscrew and remove your existing non-water-saving tap aerator and screw on your new one.

Here are some suggestions to help you reduce water consumption in specific areas around the home:

- **Bathroom:** You know the drill — install water-efficient devices and turn off those taps! For extra tips about how to be wise with water in the bathroom, refer to Chapter 2.
- **Kitchen:** Focus on how you use water in the kitchen.
  - At the sink, you can waste up to 7 litres of water a minute by rinsing things under the tap. Put the plug in the sink before washing fruit and vegies and rinsing dishes.
  - Turn the dishwasher on only when full. Better still; reduce your use of the dishwasher by washing up some of the smaller plates, knives and forks in the sink.
  - Use the dishwasher's economy cycle whenever possible.
- **Laundry:** A front-loading washing machine uses 60 per cent less water than a top-loading washing machine. You can also conserve water by washing full loads less often, rather than washing smaller loads more often. For more information about saving water in the laundry, check out Chapter 2.
- **Garden:** Put away your hose. Watering your lawn, hosing your pathways and raising exotic flowers consume lots of water. In arid climates like Australia, grass lawns are water hogs. Historically, lawns have accounted for up to 90 per cent of water used in Australian gardens, according to Sydney Water Corporation. For more information about maintaining your garden, refer to Chapter 3.

  If you have a swimming pool in your backyard, you can reduce water evaporation and avoid constant topping up by covering the pool when it's not in use — hey, this is more practical than banning the kids from doing water bombs in the pool or splashing water over the sides, don't you think? Many companies supply custom-made pool covers, but more popular are solar pool covers, or blankets, that can raise the temperature of the water in the pool by 8 degrees Celsius so that you can swim in it more often.

For guidelines about restrictions, and conserving water inside and outside the home in the area you live in, check with your local water supply agency. Table 1-1 lists key online resources across Australia. You can also get more water-saving tips at the Savewater! Alliance Web site (www.savewater.com.au).

Table 1-1 Australian Water Supply Resources

Water Supply Agency	*Web Address*
ACTEW Corporation, ACT	www.actew.com.au
Barwon Water, Vic.	www.barwonwater.vic.gov.au
Brisbane Water, Qld	www.brisbane.qld.gov.au
Central Highlands Water, Vic.	www.chw.net.au
Coliban Water, Vic.	www.coliban.vic.gov.au
Esk Water, Tas.	www.eskwater.com.au
Gippsland Water, Vic.	www.gippswater.com.au
Gold Coast Water, Qld	www.goldcoast.qld.gov.au/gcwater
Goulburn Valley Water, Vic.	www.gvwater.vic.gov.au
Grampians Wimmera and Mallee Water, Vic.	www.gwmwater.org.au
Hunter Water Corporation, NSW	www.hunterwater.com.au
Lower Murray Water, Vic.	www.srwa.org.au
Melbourne Water Corporation, Vic.	www.melbournewater.com.au
NQ Water, Qld	www.nqwater.com.au
Power and Water Corporation, NT	www.powerwater.com.au
SA Water, SA	www.sawater.com.au
SEQWater Corporation, Qld	www.seqwater.com.au
South East Water, Vic.	www.southeastwater.com.au
Sydney Water, NSW	www.sydneywater.com.au
United Water, SA and Vic.	www.uwi.com.au
Water Corporation, WA	www.watercorporation.com.au
Yarra Valley Water, Vic.	www.yvw.com.au

Tanking it

Almost as effective as curbing how much water you use is collecting your own water for free, from the sky. A *rainwater tank*, which you hook up to your guttering to collect water run-off from the roof, can pay for itself within a few years (if it rains). Most state governments in Australia offer rebates on rainwater tank purchase and installation. In fact, with a rebate, a small tank can effectively cost you next to nothing to set up.

Today's rainwater tanks are more sophisticated than the round corrugated icon tanks of yesteryear. These days, water tank systems include a range of options, such as first-flush filters to wash away leaves and debris that would otherwise enter the tank; and valves that enable you to switch between using either mains water and tank water to supply the toilet or to supply the hot-water system. Rainwater tanks come in a range of shapes and sizes, which means you can almost certainly find a shape and size that meets your particular requirements for installation. For example, they can be

- Integrated into the walls of the house.
- Stashed away under the floor (which also helps to cool and heat your home).
- Installed in the garden, as shown in Figure 1-1.

In rural areas, tank water is the preferred source of drinking water — an extra tap is fitted in the kitchen. The water pumped from rivers and dams and supplied as tap water is often too muddy and loaded with nutrients and other contaminants to be safe for drinking (although you're usually safe enough using it for washing dishes, and in the bathroom and laundry).

If you intend to use a rainwater tank for drinking water only, you need a tank with a capacity of somewhere between 400 and 1,000 litres (depending on the size of your family), which range in cost from $500 to $800. Double the size if you want to use the tank to also supply water for your garden. You also need to install a filter to make sure you're drinking clean water.

To minimise electricity, installation and repair costs, install the tank in a position higher than your kitchen sink. That way, gravity delivers the water from the tank to your drinking glass with no need for a pump.

Figure 1-1: Rainwater harvesting: This rainwater tank is designed to fit under the eaves in a confined outdoor space in the garden.

How grey is your water?

Why not reuse and recycle water that would otherwise be washed down the drain to water your garden? Governments do it and, thanks to improvements in technology and regulations, you can do it too, at home.

Water agencies define greywater as the wastewater from showers, baths, sinks, laundry tubs, washing machines, dishwashers and kitchen sinks (but not toilets — this water is called *black water*). The average 3.5-person Sydney household produces approximately 400 litres of greywater each day, which you can decide to save by easily diverting it to the garden. This approach saves you from drawing on water stored in a rainwater tank, or turning on a tap and using fresh drinking water from the mains water supply.

To capture the greywater, you can install either a greywater diversion system or a greywater treatment system. Some of the more sophisticated systems treat the water so well that you can reuse the water again in your toilet or even the washing machine.

Greywater diversion systems

In a *greywater diversion system*, water runs directly from the house through pipes and into the garden.

A basic *direct diversion system* uses gravity. The greywater simply runs down the pipes into the garden. The water flow is controlled by a greywater tap or a valve, and the flow is directed below the soil to an irrigation system within your garden. You can purchase a simple greywater diversion system for around $55.00 at a hardware store and install it yourself. Usually, installing one of these systems doesn't require council approval.

You can also get a *pumped diversion system*, which includes a tank that holds the water to control the flow and avoid possible flooding. You need a plumber and electrician to help install a pumped diversion system to ensure you meet local regulations. And, because the pump uses energy at the same time and costs money to install and run, a direct diversion (gravity flow) system is easier, cheaper and more sustainable.

Greywater diversion systems should connect to only your least polluted greywater sources, such as the laundry tub or your bath. (Other greywater sources may not be so good for the health of your garden.)

Domestic greywater treatment systems

Greywater treatment systems remove much of the soap and other sediment that exists in greywater, avoiding problems that can occur with bacteria and the build-up of waterproof sediments in the soil. You need to employ a plumber and an electrician, and get council approval before installing a greywater treatment system.

These systems are popular in sustainably designed apartment complexes, because the large set-up cost can be shared between all apartment owners. Pipes connect all greywater sources to the collection and treatment tanks, as shown in Figure 1-2. Different systems use different filters though, such as sand and soil filters, and different methods of treatment; for example, some use aeration and some use disinfectants.

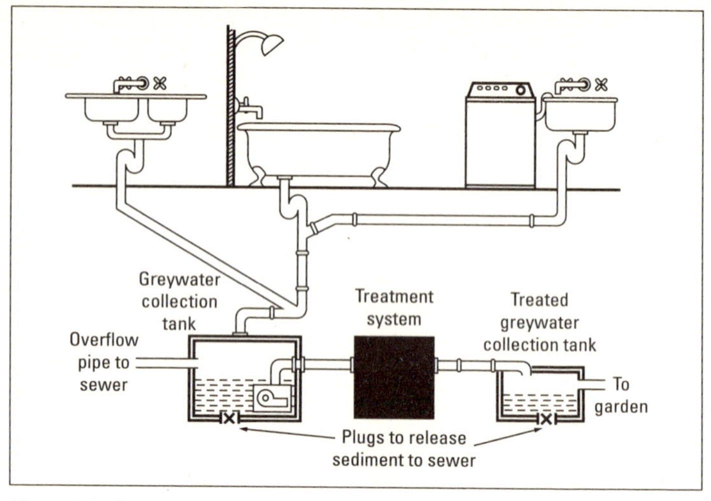

Figure 1-2: A greywater treatment system.

Drinking the Stuff

A book about water-saving tips would be incomplete without a comparison of the two main types of water that people drink: Tap water and bottled water. Yes, I thought you might be wondering . . .

How safe is tap water?

Despite treatment processes that remove harmful *pathogens* (bacteria, viruses and protozoans) from tap water, outbreaks in New South Wales of cryptosporidium and giardia in 1998 led to some serious gastroenteritis cases in the community and raised concerns among many people about the quality of the water supply.

All government agencies and water supply agencies say that there's nothing wrong with drinking tap water. In fact, they recommend it because fluoride is added to most water supplies, which helps protects teeth, especially in children. On the other hand, the same agencies don't advise drinking the water from a rainwater tank in city areas, mainly because of potential pollution and contamination problems. Country folks can, though (for more information, refer to 'Tanking it' earlier in this chapter).

Can you reduce your water bill by not drinking tap water? Not really. The water you drink is a minor component of the water you use. Finding alternative sources for drinking water isn't going to make much difference to the amount of water you take from the mains supply.

If you're worried about the quality of tap water, yet baulk at the cost of buying bottled water, you can invest in a portable jug-type of water filter (at an initial cost of around $30) or a system that is attached permanently, either on top of the bench (around $70) or under the sink (around $170 plus installation). These devices filter out the minerals contained in water, making the finished product a cleaner, clearer and more natural-tasting product.

In Australia, tap water does vary in taste, look and quality from area to area, but as long as your regular water supply meets the Australian Drinking Water Guidelines, then you're safe to drink it. You can check out the latest guidelines at the National Health and Medical Research Council Web site `www.nhmrc.gov.au`.

Bottled water is 90 per cent energy

Why do so many people buy bottled water? With tap water selling for a fraction of a cent a litre, why are people worldwide drinking more than 150 billion litres of bottled water each year?

Here's why:

- **Health reasons:** Drink manufacturers market bottled water to the health conscious as a more pure source of water (fewer chemicals and minerals).
- **Water quality and safety:** Some people harbour fears that waterborne diseases could enter the water system, so they prefer the bottled stuff.
- **Taste:** Some people prefer the tasteless and consistent purity of bottled water.

The extent of the environmental impact of having tap water delivered to your home pales in comparison to the impact of the processes involved in producing bottled water. Bottled water is the more unsustainable of the two because

- Bottled water requires plastic bottle production. Many of these bottles can't be recycled in some areas.
- Some popular bottled water brands are imported from overseas, increasing the transport costs required to get these products to you.
- Manufacturers use between 3 and 5 litres of water to make a 1-litre bottle of water. And, some estimates are many times higher.

All of these factors make the cost of bottled water around 1,000 times more expensive than a similar amount of tap water. Also, bottled water offers no relative health benefits. In fact, some commentators argue that the lack of fluoride in bottled water is responsible for declining dental hygiene standards.

Australia, You're Baking In It — Adapting to a Drier Climate

In Australia, governments at all levels are scrambling to come up with solutions to address water shortages. Population growth is outstripping available resources, while factors such as drought and climate change are reducing those same resources. Building more dams isn't a solution when not enough rain falls to fill them.

As a voter, you have a say in how government plans to supply your water or treat your waste. Understanding the issues, and participating in these debates, empowers you politically and makes you a part of the broader sustainable living solution.

Recycling urban wastewater

Governments and water supply agencies recycle wastewater in sewage treatment plants. This wastewater is now of such high quality it's considered to be an additional water source, rather than a disposal problem.

Major cities in Australia, Europe and the United States already drink recycled water mixed with fresh water drawn from rivers. Sewage is treated and returned to the river at one town, and then the mix of river water and treated effluent is re-treated and supplied as drinking water in the next town downstream. Adelaide uses this approach to treat water from the Murray River.

Large projects that pump recycled water from an urban area for use by industry are also in the works. Smaller scale solutions are possible within urban areas as well. In these solutions, recycled water is used in the bathroom, laundry and garden and is supplied separately from fresh tap water. This requires placing a third pipe alongside the water supply and the sewerage system pipes, and is practical only in new residential estates (or in large industrial and recreation areas).

The United Nations Environment Program estimates that in the next decade, one-quarter of the world's major cities are set to incorporate recycled wastewater into their drinking supply. A handful of cities in Japan, the Middle East and Europe already recycle effluent in their water supply. However, in 2006, Toowoomba residents, in Queensland, voted not to use recycled water to supplement their diminishing water supplies. Residents of Toowoomba now face the prospect of paying trucks to deliver water to their town.

Removing the salt

Desalination plants treat sea water and turn it into drinking water. Although the sea may seem to offer a potentially unlimited supply of water, the technology used in desalination plants still needs to overcome these hurdles:

- **High energy use:** The process of desalinating water is energy-intensive (and emits greenhouse gases). Powering the process with solar energy may be one way to address this problem.
- **High cost:** The cost of desalinating water is currently prohibitive for many governments, although as technologies improve these costs will become lower.
- **Environmental impact:** The disposal of the salty brine byproduct has its own environmental impact. Where and how to dispose of brine without harming the environment remains an unsolved problem as yet. The best solution for disposal will depend on where the desalination plant is located. If located on the coast, the salty brine can be disposed of in the ocean. If inland, dedicated brine ponds would be required, which has major environment and cost implications.

These problems are very real and alarm ordinary people. Western Australia gets a significant proportion of its water supply from desalination plants — because that region has no alternative. In late 2006, the New South Wales government decided to go ahead with its plan to build a desalination plant in Sydney — in response to a then worsening water supply problem — even though residents had previously protested, citing costs and the potential greenhouse gas emissions the plant would produce.

Coping with Agricultural Water Shortages

Farmers face serious difficulties coping with a decreasing water supply. Rural areas are affected by reduced rainfall, changing river systems due to the construction of new dams and reservoirs and, the big one — the siltation of rivers and streams.

Impact on agriculture

Because 80 per cent of Australia's water is used by agriculture, any reduction in the water available for irrigation purposes affects Australia's economy. Whether or not this water can be used more effectively is a hotly debated topic, but certainly rural Australia faces a serious water crisis.

Programs underway to make more efficient use of the water available include

- Reducing irrigated water allocation to levels that match the amount of water inflow from rainfall (this is now the case in the Murray–Darling Basin).
- Concreting and covering existing irrigation channels, or replacing them with pipes to reduce losses due to evaporation and leakage.
- Understanding how much water each crop actually requires — to help farmers use only the water they need.
- Introducing drought-tolerant species, especially wheat, in dry areas, and more intensive farming practices in high rainfall areas.

Some of these efficiency innovations, however, create new problems.

- New species created by *genetic modification* (altering the biological characteristics of a food crop by using genetic material and proteins from another source) lead to major concerns about the health impacts of eating genetically modified food. The process is controversial because it introduces new allergens, toxins, disruptive chemicals, soil-polluting ingredients, mutated species and unknown protein combinations into people's bodies and the wider environment.
- Intense land use in high rainfall areas may increase silt and pollution in nearby waterways, creating new areas of non-productive land.

Impact on remote communities

Communities not connected to mains water supply usually get their water from nearby groundwater, rainwater tanks or a combination of both.

Groundwater is the most reliable source of water in central and southern inland Australia. Some of these infrastructures are poor, though, which compromise water quality. The good news is that upgrades are in the works in many regional areas of Australia.

Many rural properties also sink their own bores to reach groundwater. This process involves boring to reach the underground water table, then pumping the water to the surface and storing it in a tank.

When remote communities really run short on water, short-term remedies include transporting water from reservoirs.

Chapter 2

Saving Water in the Home

In This Chapter

- Reducing your water use in the bathroom
- Taking care of water business in the kitchen
- Looking at laundry water-saving options
- Going solar

Tired of keeping up with the Joneses? Are you more interested in making your home a more water-efficient environment?

In this chapter I get to the heart of lifestyle matters — the way you live with water at home. I take you through the main rooms in which the most water is used and show you how each space provides you with some unique opportunities to use less water. Along the way, I offer a range of money-saving and eco-friendly ways to manage water in your home, most of which are simple ideas that you can start using today.

Bathroom Blitz

In the bathroom, you can wash your worries away. You can spend some quiet time by yourself, relaxing in your bath or showering away a day's worth of sweat and tears. Reducing your consumption of water greatly reduces the demands you place on the urban supply and helps you cut your water bill. And, everyone's daily bathroom routines can also help reduce the water stress placed on the planet.

Most unnecessary home water waste occurs in the bathroom. Here are some guidelines to help you reduce the amount of water that may be pouring down your bathroom's drains:

- **Don't let taps run:** Some commentators say that if you allow the water to run for three minutes while you clean your teeth, wash your face or scrub your hands, approximately 15 litres of water goes down the gurgler. Another point: You're wasting a lot of good water if you allow water to run down the drain while you're waiting for it to heat up. If your plumber can't fix this 'hot water delivery problem', put the plug in the sink or catch some of the water in a bucket and use it to water your plants.
- **Reduce your showering time:** Sure, I know it can be difficult to take short showers, but if you set a daily time limit on your showering time, to maybe five minutes, you develop a routine that you can more easily stick to.

  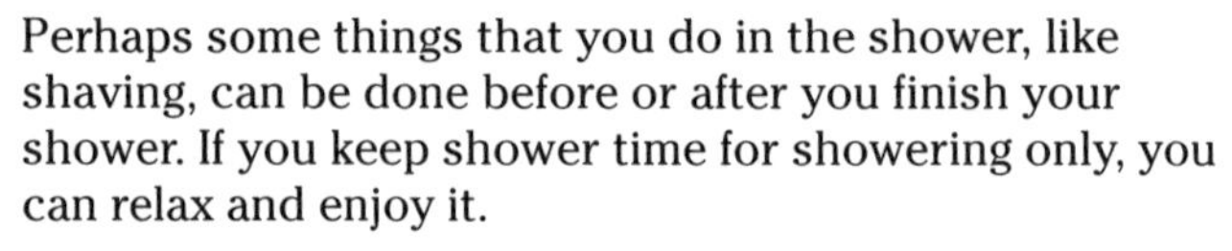
  Perhaps some things that you do in the shower, like shaving, can be done before or after you finish your shower. If you keep shower time for showering only, you can relax and enjoy it.

  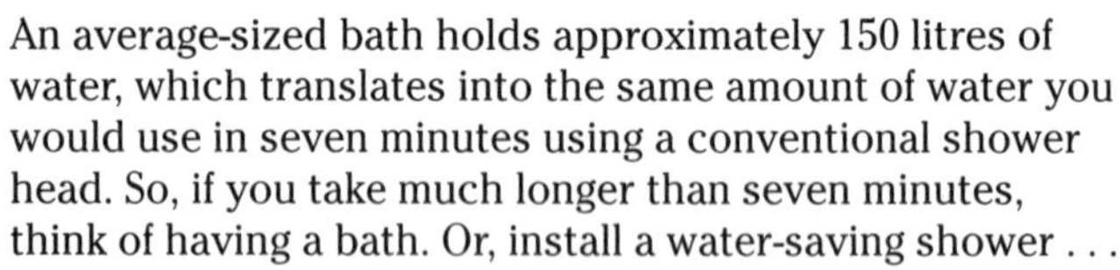
  An average-sized bath holds approximately 150 litres of water, which translates into the same amount of water you would use in seven minutes using a conventional shower head. So, if you take much longer than seven minutes, think of having a bath. Or, install a water-saving shower . . .
- **Get water-saving shower heads and taps:** Many new homes are already installed with water-efficient AAA shower heads and taps. But if you live in an older home (that hasn't been renovated recently), call in a plumber and ask him to check your shower heads and taps, and replace them if they're not up to sustainable standards. A new water-saving shower head can save up to half the amount of water that you normally use showering under an older shower head. You also need to heat less water.
- **Press the correct button on the toilet:** These days, most toilets have half-flush and full-flush options (called *dual-flush* toilets in most homeware stores). Use the half-flush when you can. Also, ensure that your toilet doesn't overflow and leak — a major source of water waste in the bathroom. Toilet leaks are usually a result of an incorrect setting in the cistern, which a plumber can fix in a couple of minutes.

- **Install a toilet suite with an integrated hand basin:** This innovative combo can reduce total bathroom water usage. Check out Caroma's version (which is also 5-star rated) at www.caroma.com.au.

The Water Efficiency Labelling and Standards (WELS) scheme, shown in Figure 2-1, rates the water efficiency of taps, shower heads and toilet suites. This compulsory scheme ensures manufacturers submit their products for water conservation rating and display this label on their wares. For more information, visit the WELS Web site at www.waterrating.gov.au.

The waterless toilet

Of course, not every household can have one, but the waterless composting toilet is becoming popular in sustainably designed homes.

Similar to the backyard septic-tank *dunny,* this modern-day variation can be installed within the home in the bathroom, with the waste collected in a container or chamber below the house for treatment to create compost for fertilising your garden. Unlike the smelly old backyard *dunny,* the modern composting toilet smells no worse and looks very similar to the flushing variety you're already familiar with.

Two types of composting toilets are on the market: The continuous composting toilet, which decomposes as it moves slowly through a composting chamber; and the batch composting toilet, where the waste is collected in a container and moved away to compost separately. You usually need at least two containers for a batch system so that you can alternate when you move a batch of waste away to compost separately.

There is no doubt that waterless composting toilets require more attention (and can be expensive to switch to — from between $2,000 and $5,000), especially at the composting end of the process, but the water that you save is significant.

Several companies in Australia sell off-the-shelf composting toilets. Check the following manufacturers for more details:

- Rota-Loo: www.rotaloo.com
- Nature Loo: www.nature-loo.com.au
- Clivus Multrum: www.clivusmultrum.com.au
- Biolytix: www.biolytix.com

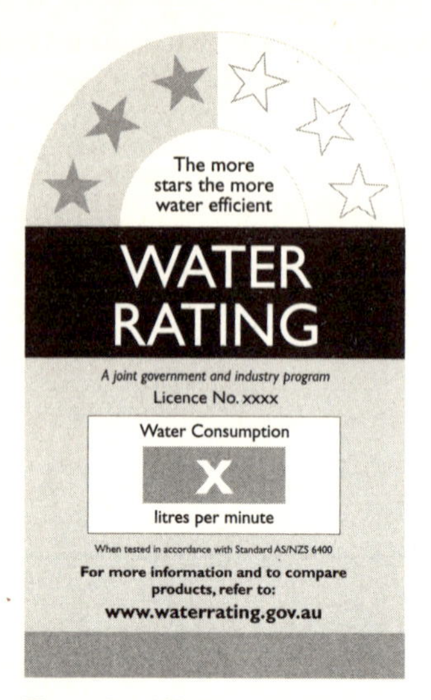

Figure 2-1: Water-rating labels on new products indicate how water-efficient they are. The more stars highlighted, the better.

Kitchen Connection

The kitchen is said to be the social soul of the home but how you work in this room — prepare meals and clean up — can have a major bearing on your water usage.

Here are some tips for reducing the amount of water you use in the kitchen:

- Rinse and prepare food in a sink half-filled with water rather than under a running tap.
- Wash some of your dirty plates, knives and forks in the sink the old-fashioned way to reduce your reliance on the dishwasher.
- If you're washing the dishes by hand, don't wait for water to get hot before putting the plug in the sink — put the plug in straight away so as you capture all the water that comes out of the tap.
- If you need to have a dishwasher (remembering that the most water-efficient dishwasher is you), buy one that has a high water-efficiency label — the more stars the better.

- Before putting your dirty plates in the dishwasher, wipe or scrape food scraps off rather than rinsing them off under a running tap.
- Use the dishwasher only when you have a full load instead of doing a wash after every small meal.

Laundry Time: Doing an Eco-Wash

The laundry is often the least-liked room in the home — it usually reeks of chores and hard work. Thanks to good engineering, however, washing machines are easy to use — you can simply load 'em, then set and forget. A washing machine can use a lot of water, though, especially if you treat it as the workhorse in an above-average sized family.

Calculate how many times a week you put your washing machine through its paces, then use the following tips to begin water-saving practices:

- **Wash less often:** Just like the dishwasher, use the washing machine only when you have a full load ready to wash. And put only those clothes in the washing basket that are really ready to wash. You can also reduce the amount of washing you do by individually soaking or hand-washing more troublesome items in hot water in a bucket.
- **Get a front-loading washing machine:** Front loaders use less water than top-loading washing machines. Some states in Australia even offer government rebates if you buy one of these water-saving washing machines.
- **Reuse the wastewater:** The greywater waste generated by your washing machine (and also your shower, bath and sinks) can be used to water your garden. For more details about how to capture your greywater and pump it out to the garden to keep your plants alive, refer to Chapter 1.
- **Use eco-friendly washing detergents:** Unlike regular commercial brands, eco-friendly washing detergents don't contain phosphates or petrochemicals, so they don't harm the environment when they enter the water system — which means you can safely use the greywater from your washing machine for watering your garden. Look for earth-friendly products in your supermarket.

Using the Sun to Heat Your Water

Although installing a solar heater does nothing to reduce the amount of hot water you use, many people who set up their bathrooms, laundries and kitchens to be more water efficient inevitably look at the cost versus the benefit of installing solar heaters.

A solar water heater can provide between 50 per cent and 90 per cent of your total hot water requirements, depending on the climate in your area and the model of heater you buy.

The basic principle is simple: Cold water is spread out over the roof in pipes or glass tubes and is heated by the sun. The water is then pumped, or moved by its own heat, into a holding tank while more cold water is heated up. On a sunny day the water quickly reaches boiling point, so all systems are fitted with safety devices to stop the water from getting so hot that it burns people.

Several types of solar heaters are on the market, all with different characteristics and varying degrees of energy efficiency. Here is a run-down of the different solar heating models you may encounter:

- **Passive water-collection heaters:** Cold water is collected in a tank in the roof, which is heated during the day by solar panel plates. Passive systems are the most energy-efficient solar heater because they don't use electricity to pump the water. You can choose between two types:
  - Open circuit system, where water flows directly from one tank to another.
  - Closed circuit system, where a heating fluid is warmed in the tank and then the heat from it is transferred to the water already sitting in the collection tank. The closed circuit system is commonly used in areas that experience frosts or freezing conditions in the winter. The fluid used to generate heat for the water in a closed-circuit system has anti-freeze properties.

Passive systems use either gravity fed or closed-coupled flow. The gravity-fed model has the water collector situated above the storage tank in the roof, and normal gravity moves the water through pipes from collection to storage and through to your taps. The closed-coupled system has the collector below the storage tank and uses normal water pressure to get the water to storage and through to your taps.

- **Active water-collection heaters:** If you have limited space in the roof or a traditional hot water system, you may go for an active rather than a passive system. The storage tank on an active system can be located on the ground or under the house, with the heated water pumped from the collector tank to the storage tank.

  Active pump systems, such as the one shown in Figure 2-2, are good for converting your existing tank to a solar powered system, but you need to power the pump somehow, and the obvious way, unfortunately, is with electricity.

Figure 2-2: An active solar hot water system.

No matter which type of solar power system you choose, be sure to position the solar panel in the roof at an angle that takes advantage of the sun's rays. Also, think about what size tank you need to cater for your family; try not to buy something larger than you really need.

Whether you choose a passive or active solar water-heating system, if the sun's not shining, your solar power system can't heat the collected water. To overcome this, you can buy a booster to heat the water when the temperature falls below a certain point. The boosters can be powered by electricity, gas, petroleum or diesel.

However, boosters can waste energy if they're used too often. In fact, some boosters require more energy than normal off-peak electrical hot water systems. Your best bet is to install timers if you find your booster system is being used too much.

To find out more about solar water-heating systems, including buying a suitable unit for your home, check out the Australian Greenhouse Office's Technical Manual for Environmentally Sustainable Homes at `www.greenhouse.gov.au/yourhome/technical/fs43.htm`.

Just love that piping hot water

According to the Australian Greenhouse Office (AGO), in most households hot water is the largest energy cost and cause of greenhouse gas emissions. In fact, heating water accounts for about 27 per cent of an average household's total greenhouse gas emissions and energy use.

This practice is an incredible luxury, especially when you consider the low cost of getting the sun to do it for you. If you can't afford to install a solar hot water service, you can save water and energy by installing water-efficient fittings and appliances in the home.

Chapter 3

Greening Up Your Garden the Water-Wise Way

In This Chapter

- Designing and maintaining a water-efficient garden
- Becoming a compost convert
- Going native
- Taking a look at your lawn

You'd think it would be easy being green in the garden. All that lawn, and all those flowers, tree and shrubs . . . they make many homes look like shrines to the natural environment.

Appearances can be misleading, though. A well-manicured garden can drain water and other resources from the environment. Take a moment to think about it. A constant supply of water is required to keep poorly designed gardens alive and green. The usual tools employed to keep a garden neat and tidy are motorised lawnmowers and other power tools. And toxic chemicals promise to keep pests at bay and feed the soil and roots.

With some direction and guidance, you can turn a heavy reliance on water around and create a garden that makes a resoundingly positive impact on the environment. In a sustainable garden, you grow and eat your own organic produce; you recycle your food scraps to feed and replenish your plants; you create your own compost; and you add value to the local ecosystem by growing native and indigenous plants. And, all of these methods help to reduce water usage.

Even if you live in an apartment block, you can use pots to bring nature into your life and enjoy the benefits of cool green leaves, colourful flowers and growing your own fresh food or herbs.

In this chapter, I help you create your own sustainable garden. I show you how to design and plant your own fruit trees and vegetable patch, and set up a composting system. I also introduce you to the types of plants and shrubs that complement this approach. In particular, I encourage you to plant native Australian species, because they need less watering, encourage indigenous birds and animals, and require less effort to maintain.

Finally, I deal with the lawn. You probably already know that large tracts of grass require extensive watering, weeding and mowing to look immaculate and evergreen. Nevertheless, a lawn provides a great space for playing or to hold a party. I show you how to keep your lawn at its best without draining the planet's precious water resource.

By the way, the advice in this chapter only scratches the surface, so to speak, of gardening. For more good gardening techniques, check out *Gardening For Dummies*, Australian & New Zealand Edition (published by Wiley Publishing Australia Pty Ltd).

Designing Your Water-Wise Garden

Everyone can have a garden. Some people keep a small, potted garden on an apartment balcony, or have a plot in a community garden that belongs to the whole neighbourhood. Others have more space to work with and can maintain a traditional suburban backyard garden, or run a large market garden that feeds the local community.

Whatever the type of space you have, the objectives of creating a sustainable and water-wise garden are the same:

- Minimise the amount of water and chemicals you use.
- Grow plants and vegies to reduce your impact on the environment.
- Create your own plant food through recycling your waste. (Compost is a critical component of any sustainable garden. I explain how to get compost started in the section 'Composting Your Waste' later in this chapter.)

Planning the landscape

Before you go out the back and start digging soil and planting seeds, step back and think about what you want. You need to work out what's achievable in the space that you have and its geographic and climatic limitations. Most importantly, you need to design a garden that you're going to enjoy maintaining and using.

To begin, map out a rough design or landscape plan. Sit down, take a deep breath, and on paper sketch out how you'd like to fill the space. Here are some suggestions to help you get started:

- Identify the things you want to keep in your current space — for example, the clothesline, the barbecue and seating area — and what you want to move or downsize. Do this even for a balcony garden. Your design needs to take into account all the activities that take place in the area.
- Plan where you want to put pavers, pergolas and other functional landscaping features such as a bird bath.
- Break the space into different functional areas — some for growing food, some as a home for local fauna and some to simply look good or provide flowers to decorate your home.
- Identify on your plan the shady and sunny areas, and the areas that are easy to water or get plenty of rainfall. (For more details about how to approach working with nature, see the section 'Working with the lay of the land' later in this chapter.)

If you're limited in space, don't be afraid to put plants close together. Many gardeners promote vertical growing — placing plants in containers on racks one above the other — in spaces that are tight. Less space also requires less soil and less watering.

If you're stuck for ideas on how to landscape your sustainable garden, check out magazines and a good gardening book, or chat to your local nursery.

Planning what to plant

After you map out the space, consider what plants you want to grow and how often you're likely to use them. For example, the kitchen garden, with herbs, leaf vegetables and other regularly used plants, is better off close to the house.

- The compost heap and the worm farm (and any animals) are better off some distance from the back door, but need to be easily accessible for the once or twice-a-day trip you need to make to reach them.
- Fruit trees and bushy herbs such as rosemary, bush basil and lavender need less attention, as do larger food crops such as corn, potatoes, eggplants and sunflowers.

  Native fruits are pest resistant and well adapted to the local climate and so require less water.
- Soft fruit crops like strawberries and tomatoes warrant regular attention, so they're better placed closer to the house — or, at least, visible from the kitchen window.

- Fruit trees and vegies usually require more sun and rain than other plants, so make sure they're out in the open — away from fences and shady areas — without being exposed to howling winds or other damaging weather that may be prevalent in your area.

  Don't forget to place vegetable gardens and fruit trees where you can get plenty of water to them. Rainwater from the roof and wastewater from the kitchen, bathroom and laundry can all be delivered by a direct diversion system. This method uses gravity to direct the flow below the soil to a simple dripwater system of irrigation to keep your plants vibrant and healthy.

(If your house is low on the block, you may need a pump; refer to Chapter 1 for more details.)

As you juggle these components around on your sketch, you're likely to discover that some elements find the 'right' spot straight away. You can start organising the rest of the garden around these items. Obviously, any building or major earthworks needs to be done before you can plant anything that can be disturbed by construction.

Working with the lay of the land

To create a sustainable and water-wise garden, you need to work with nature and minimise the amount of effort and water you use. Here are some questions to consider before you begin planting:

- **What is your local climate like?** For example, do you experience frosts, very little average rainfall, or a relatively humid climate with a good level of moisture? Plants requiring sun, and those that can cope with heat — like many fruit varieties and vegies — can be placed in more exposed areas. More sensitive varieties can be planted in shady zones or areas not exposed to prevailing winds.
- **How much area do you have to work with?** A large area provides more scope for fruit trees, whereas a small area requires you to be selective about what you can grow. A large area also needs more maintenance, so create some areas that can look after themselves.
- **What is the slope like on your property?** Does the property drain well? Or is it basically flat and therefore retains water? Plants requiring more shade and moisture are best placed near fences or near the side of the house, as well as on any downslopes and low lying land where water drains to.
- **How does the shade fall across the garden?** Keep the sunny aspect — the northern side in Australia — for plants that enjoy full sun, and put low trees that don't create too much shade in winter on the other side. Identify the areas that get shade for most of the day and some of the day, so you can plan and plant appropriately.

- **What's the soil like?** If the soil is sandy it can drain well. If the soil is thick with clay, it can't. The best growing soil has an equal mix of sand, organic matter and clay.

  Check the requirements of the plants you want to grow with your local garden centre. It may be that you need to buy some organic or other specialist soil to grow the plants you want.

- **How often does it rain?** Yes, here now is the best question for last: How are you going to water your sustainable garden? Hardy plants, like Australian natives, can look after themselves, but tomatoes require lots of water. (For more information about watering a garden, see the section 'Minding your water usage' in this chapter.)

No, that mango tree your father-in-law sent down from Cairns can't grow in a temperate climate, but an apple tree can. When you get a feel for what grows in your local conditions, you can plant accordingly.

Maintaining a Sustainable Garden

Your research identifies plants that do well in your climate, and the soil, moisture and sunlight conditions that those plants prefer. This information determines where in the garden the various plants need to go, and what you need to do to keep them hale and hearty.

Minding your water usage

Some people living in drier regions have no qualms about turning on the sprinkler and watering to their heart's content. This approach is short-lived and gets more difficult as water restrictions become a permanent feature of Australian life.

The simplest method to reduce your usage of mains water is to install water tanks to capture and store rainwater from the roof of your house and sheds. Water tanks are becoming so popular again that they're now the norm in many new housing developments. Some state governments in Australia even offer a cash rebate on rainwater tanks to encourage their use.

Wastewater from the sink, shower or washing machine is known as *greywater*, named because the soap it contains sometimes turns it that colour. Greywater needs some treatment to prevent it clogging up the pores of the soil and making it hard and water resistant. You're best to use it quickly to prevent bacteria building up in tanks and, ideally, soak it into the ground using an agricultural pipe or drip-filler system, rather than pouring on the surface. For more information about greywater systems, refer to Chapter 1.

Keeping a check on your garden's health

The key to maintaining a sustainable garden is healthy soil. Good soil reduces your need to use unnatural additives to maintain the health of your plants. Also, by designing your garden to take advantage of the natural conditions of your local ecosystem, you give yourself the best chance of success.

Here are some tips that improve your garden's health and make it easier to maintain:

- **Discourage pests:** Rotate your crops from season to season, prune in winter, harvest your crop just as your food becomes ripe and make sure you don't leave anything behind that might rot away.
- **Reduce weeds in your garden:** Weeds can attract unwanted pests that can ruin your seedlings and plants, and compete with your vegies for sunlight, water and nutrients. Nipping weeds in the bud, so to speak, before they get a hold is the best way to reduce their influence. However, don't be tempted to use toxic chemicals to kill them off quickly.

  You can also reduce the influence of weeds by adding mulch to your garden beds. Organic mulch keeps the weeds at bay and also insulates roots on hot days and retains soil moisture. Consider mulch as insulation batts for the backyard — a water-saving plus.

- **Use organic fertiliser to feed your plants:** Organic fertiliser contains rock minerals and animal manure that is produced from sustainable farming methods. You can use your own composted material, or the liquid from the bottom of a worm farm, to fertilise your plants. For more details, see the section 'Composting Your Waste' later in this chapter.
- **Use organic insecticides:** Try mixing this: Garlic, chillies and dried pyrethium. This natural insecticide doesn't poison the environment. In fact, garlic has been used as the main base for naturally made insecticides for thousands of years. The chillies and dried pyrethium add more punch (to deal with today's teenage mutant grasshoppers — just kidding).

Improving the chemistry for a healthier soil

The key to healthy soil is making sure you keep up the nutrients. You can do this quite nicely by feeding your soil with the following natural materials:

- For phosphorus upkeep, which stimulates seed and root growth, use rock phosphate or any other minerals containing phosphate.
- For nitrogen and protein upkeep, which stimulates green growth, feed compost piles with alfalfa or cottonseed meal. Grow peas or pea bushes such as pigeon pea.
- For potassium upkeep, which helps plants resist disease, use glauconite, sulfate of potash or even wood ashes.
- For calcium upkeep, which helps with root and leaf growth, use gypsum.
- For magnesium upkeep, which helps plants stay green and healthy by promoting the production of chlorophyll (the good stuff that plants use to process carbon dioxide), use dolomite lime.
- For sulfur upkeep, which helps feed the life that works away underneath the soil, use gypsum.

- For oxygen upkeep, how you garden is more important than what you add. Ensure that the compost, manure or any of the minerals you add are turned over and through the soil regularly to maintain air pockets and encourage good root growth.

After you add some of these nutrients to help your soil to be healthy, you simply need to regularly add compost to maintain its health. For more information about creating and maintaining a sustainable garden, check out the online resources listed in Table 3-1.

Table 3-1 Online Gardening Resources

Resource	*Web Address*
Introduction to sustainable garden techniques	www.gardensimply.com/technique.shtml
Developing a sustainable produce garden	www.sgaonline.org.au/info_producegardening.html
Produce gardening	www.sgaonline.org.au/info_producegardening.html
Growing fruit and vegies in containers	www.container-gardens.com
Sustainable pest control	www.greenharvest.com.au/pestcontrol/general_purpose_spray_prod.html
Getting involved in a community gardening project	www.communitygarden.org.au
Sustainable composting techniques	permaculture.org.au/?page_id=22
Setting up a worm farm	www.epa.nsw.gov.au/envirom/wormfarm.htm
Water-wise gardening	www.sgaonline.org.au/info_water_conservation_in_the_garden.html

Growing food naturally

When you research sustainable gardens, you come across a variety of techniques designed to achieve a healthy garden with a minimum of resources. These techniques each take a particular approach to growing food naturally. They include:

- **Organic gardening:** This method encourages the use of feeding and maintaining soil by using natural methods. Essentially, this approach is gardening without the use of petroleum-based products and artificial chemicals. Using organic gardening practices also conserves water.
- **Permaculture:** Developed in Australia, this technique promotes the development of your own ecosystem so that the garden maintains itself. For example, kitchen waste can feed chickens and the compost heap to fertilise your garden soil, which creates the food you eat. The wastewater from your house can be treated in ponds that support fish and plants and then used to water your plants. *Perennial* plants form an important part of a permaculture garden. (Perennial plants, such as fruit trees, are those that don't need to be dug up and replaced each year.)
- **Biodynamic gardening:** A study of life in soil and water led to this philosophy of gardening and farming. Special compost and food preparation improves the quality of the food and plants grown.

  According to the Australian Bio-Dynamic Research Institute, *biodynamics* is the process of farming to 'culture' the soil. In biodynamics, every aspect of primary production is based on the dynamic interplay between soil life, plant and animal health, and how these all benefit each other.
- **Fukuoka farming:** When is a technique not a technique? When you farm the Fukuoka way! Plant your seeds and then just leave everything to the elements, with no assistance whatsoever. Water is the clear winner!

- **Bio-intensive farming:** This technique involves growing food and plants in a very small area. You can adopt this method by planting seeds and plants in raised boxes on top of each other. Bio-intensive farming is a bit like promoting living in apartments as being more sustainable than living in low-density areas because it involves less consumption of resources. Bio-intensive farming results in you using less water and less energy (human and natural) to keep the garden healthy.

Composting Your Waste

Composting is nothing to turn up your nose at. When you realise all the benefits, you may actually be eager. Composting is a magic formula for maintaining a healthy, sustainable and water-wise garden. In fact, composting creates your own household mini-ecosystem where the waste that becomes the food, becomes the waste, becomes the food . . . in a never-ending cycle. It doesn't get any more sustainable than that!

Not only is compost essential for producing and maintaining many of the nutrients that give your garden vigour, it also helps the soil perform at its peak when your plants start growing. Here's how:

- A rich, healthy soil retains much of the moisture that it receives, which means that you don't have to overdo it when using your captured rainwater.
- Compost is full of living things that hold the nutrients, so the roots of your plants get the chance to absorb them before they leach from the soil.
- The organisms in compost keep the soil healthy by reducing soil-borne disease and the need for chemicals that try to do the same job.

Composting also gives you an opportunity to get rid of your kitchen waste sustainably. Composting is all good news.

Keeping chooks

The humble hen has been an integral part of human settlement since the dawn of agriculture. Whether you call them chickens, chooks, pullets or poultry, *Gallus domesticus* — to give them their scientific name — can produce a bounty of free-range eggs and speed up the process of recycling your kitchen waste. They're usually kept by people living in rural residential areas, on small acreages or with larger backyards.

A small hen-house, chook-shed or chicken coop in one corner of the backyard with three or four hens can keep most families in organically grown, free-range eggs. You may never need to buy an egg again. Your hens also produce an ongoing supply of nitrogen-rich fertiliser.

Let the chooks out into the garden — keeping them away from seedlings and delicious low-growing fruit — and they can keep down snails, caterpillars and grasshoppers and turn over the top layer of soil with their powerful fork-like feet.

Follow these tips to make sure that you and your chooks live happily together as a harmonious sustainable whole:

- Make sure the shed has plenty of air, but the birds aren't sleeping in a draught.
- Keep the floor of the shed dry, and clean out the manure every couple of months.
- Design the shed to keep out cats and other predators common to your region (such as pythons, if you live in the north of Australia).
- Design the water supply to last a couple of days so you can go away for the weekend.
- Remove uneaten food scraps and put them into the compost.
- Complement your hens' diet with commercial pellets, but don't be surprised if they turn up their beaks and wait for more table scraps.
- The more often you let them out into the garden, the happier they're going to be.
- Think carefully before introducing a rooster. They're noisy and aggressive — so noisy they may breach council regulations.

Rodents see chook sheds, and compost heaps, as a free supply of food. By building strong enclosures and feed containers that keep the food tidy, and by removing waste on a regular basis, you can minimise the rodent problem.

Note: Some local councils have rules about keeping chooks, such as a limit of four birds in some metropolitan areas or a 'no noisy roosters' regulation. Check with your local authority before you proceed.

Beginning an outdoor compost

You can go to your local garden supplier to buy compost — but you probably don't need to. You very likely produce enough varied waste in your kitchen, which can be mixed together to form your own compost. Here's what to do:

1. **Get a sustainably made compost bin (or make one of your own).**

   You can buy a nice, sustainably made outdoor compost bin at most plant centres or hardware stores. And if you live in a city without a yard, consider an indoor worm farm or a composting system such as the Bokashi bin that enables you to recycle food scraps without the mess of smells.

2. **Get your ingredients together.**

   The key to good compost is balance — the organisms that break down your food scraps work best when mixed with water, air and a good balance of leaves. Don't overdo it with any one ingredient, though, or you may end up with a stinking pile of rotten food.

   You don't have to go too far to find the two key ingredients of compost: Carbon and nitrogen. Take a walk to your household paper recycling bin to find some carbon (paper, bark and sawdust, for example), then go to your kitchen bin to find nitrogen (fruit peels and vegetable wastes). Add some grass clippings, manure (trot down to your local stables for some) and weeds to the mix and you're on the way to having some great compost on your hands.

   Equal proportions of each ingredient in the mix typically do the trick, as long as everything is mixed, watered and aerated well.

3. **Maintain your compost by keeping the mix moist, but not wet.**

   As you're adding material to your compost, make sure you keep the heap of festering compost moist. Water and air allow the thousands of bacteria and fungi to do their work. Too much moisture and you create some of the most pungent smells your neighbours have the pleasure of whiffing. Too dry, on the other hand, and nothing happens.

The key to keeping air available for these micro-organisms is to turn the compost over regularly (for example, once a week) with a garden fork and allow as much oxygen to enter the process as you can.

4. **Harvest the compost.**

   A good compost brew looks like dark soil and contains no recognisable food scraps. (This process can take anywhere between two and ten weeks, depending on local conditions and the mixture of ingredients.) You can then scoop out what you need and dig it into your garden.

The composting that I just helped you set up can be defined as *warm composting* — anyone who has lifted the lid on a composting bin can appreciate this. In fact, the organisms that break down the food can produce so much heat that grass clippings can turn to ash. So, make sure your compost stays moist.

Setting up a worm farm

Worm farming is *cold composting*. A worm farm makes a great composting alternative if you don't have a backyard (or use an inside composting system).

Worm poo! Hard to imagine, but a worm farm is one of the most effective composts you can create. It can all take place within a confined cool space, like a laundry or a well-shaded balcony. Or your worm farm can sit quietly on a back porch or on an inside bench and produce nutrient-rich juices and humus, without many of the smells associated with regular compost bins.

Becoming a worm farmer is easy. The following information can help you get started:

- You can buy a worm farm or box from your local council office (check your council's Web site to find out whether it stocks them), or make your own worm farm from scratch.

  To make your own, get hold of some storage boxes or crates that aren't being used for anything else — you need four for maximum effect. The bottom container needs to be waterproof and large enough to take the weight of the other boxes when they're full of soil.

- Many of the worm farms you can buy have all the necessary start-up material contained in the box, from the right types of worms down to the soil. Most experts recommend two types of worms: Red worms or tiger worms. Don't bother with your average garden worm — they don't do the job, so to speak.
- For a worm farm to work, you need to have a good supply of vegetable scrap waste. Also, make sure that a sufficient number of holes, or perforations, are in the bottom of each box (except for the one that sits at the base) for the worms to move from box to box. Start with three boxes plus a waterproof container at the bottom to collect the worm juice.

Armed with the preceding information, follow these steps to get your worm farm underway:

1. **Set up box one.**

   Line the bottom of the first box with soil and newspaper, add some fruit and vegie scraps, then add the worms and block as much light to the box as you can by placing a Hessian cloth or some more newspaper over the top. Place this box on top of the waterproof container.

2. **Feed and nurture your worms in box one.**

   Worms don't like acidic food, so don't put orange and lemon skins in the worm farm. Raw onion, tomatoes and pineapples aren't good for them, either. Remove the newspaper or Hessian each time you add food scraps and replace it afterwards. Add leaves, paper or other bulk with every second or third batch of food scraps. Spray the contents with water every now and then to keep it moist.

   After a couple of weeks of monitoring, the worms in box one grow larger and multiply and the box becomes full.

3. **When box one is full, set up box two, then later, box three.**

   Set up box two the same way you set up box one (in Step 1). Then remove the Hessian or newspaper from the top of box one, and place box two on top. Keep adding scraps to box two, the same way you did for box one. When box two is full, do the same with box three.

4. **Harvest your worm compost material in box one.**

   By the time box three is full of food scraps, the worms have finished eating all the food in box one and moved on. Lo and behold, all that is left in box one is an accumulation of compost material that you can spread over your garden or through your pot plants.

5. **Harvest your worm juice.**

   Now that your worm farm is well established, the waterproof container at the bottom of the stack regularly fills with liquid fertiliser (don't worry, it doesn't stink). Dilute this worm juice — two-parts water to one-part juice — then pour this mix onto the plants you want to fertilise.

Choosing and Planting Native Shrubs and Plants

Growing Australian native plants is good for the environment. They're also easy on you: Native plants and shrubs don't need much looking after.

Natives don't require as much watering as imported species because they're adapted to the generally arid Australian conditions. They also provide food for bird and insect species, ensuring your garden is the hub of an active community.

Another environmental advantage is that you're helping preserve the local plant stock. Humanity's spread across the planet has been responsible for replacing rich, diverse plant communities with a small number of favourite flowers and fruits. You can reverse the trend by buying a local plant from a local nursery.

Not only should plants be native to your country, they thrive if they're also indigenous to the local area you live in. Indigenous plants are adapted to local conditions and interact with the local ecosystem more effectively than other species. For example, the plants indigenous to the Snowy Mountains area in New South Wales are different to plants indigenous to Queensland's tropical north. Indigenous plants in both of these regions differ from one another because they've evolved to work best with the average rainfall and heat in their own regions, as well as the local birdlife, insects and soil conditions.

Listing all the different regional conditions that exist around Australia, and the types of shrubs and plants that should be planted in each area just isn't possible. Instead, here's a list of the main types of indigenous plants that you can use as a guide:

- **Wattles (acacias):** Most acacia plants can be grown in many regions of Australia, especially coastal, mountain and arid inland areas.
- **Banksias:** Most of the banksia species originated in the southern region of Western Australia but are diverse enough to be grown in most regions of Australia.
- **Boronias:** A variety of boronia species can be found throughout Australia, although they rarely flourish in humid or arid areas.
- **Bottlebrushes (or callistemon):** Bottlebrushes work well in heavy rainfall areas and along waterways in forested areas.

- **Dyandras:** Dyandra is one of the many species that occur naturally in south-western Western Australia. They seem to work best in areas of dry summers and wet winters and in sandy soils.
- **Eucalypts:** This species forms much of the Australian bush and is commonly called a gum tree. Eucalypts grow just about anywhere except in rainforests.
- **Grevilleas:** This is another native that grows just about anywhere. The flowering capabilities of the grevillea make it one of the more popular garden plants.
- **Melaleuca (or paperbarks and honey myrtles):** These plants are generally found along watercourses in woodlands and shrublands all over Australia. They're popular landscaping plants that require moist conditions to flourish.

For some online resources that cover more indigenous plant species, check out Table 3-2.

Table 3-2 Sources of Native Plants in Australia

Type	Web Address
How to choose the best plants	www.sgaonline.org.au/info_live_local_plant_local
List of native species nurseries	www.greeningaustralia.org.au/GA/NAT/TipsAndTools/Nurseries.htm
List of indigenous plants for different regions	www.floraforfauna.com.au
List of native species	http://asgap.org.au/sgapla.html

Living With a Lawn

Millions of suburban gardens are little more than a lawn surrounded by a few shrubs. Keeping the lawn lean and green takes up all the time many people set aside for backyard chores.

Much of the time you spend on manicuring your lawn is probably unsustainable. Here are the main reasons why:

- The amount of water you need to keep the lawn green and healthy is wasteful. Mown grass is a water hog because it doesn't produce shade and has relatively shallow roots.
- Using petrol-powered lawnmowers and other cutting devices to keep things trim is an unsustainable practice. That Sunday sound of suburban lawnmowers keeping the grass at a tidy height and the edges nice and clean means a large amount of fuel is being used, adding to greenhouse gases.
- Chemicals used to rid the lawn of weeds and pests and to keep the lawn green are pretty toxic.

One key consideration in sustainable garden design is to reduce the amount of lawn and to use that space as part of a mini-ecosystem that has its own fruit and vegie garden, water collection system and compost area.

To get more out of your lawn, try these hints:

- Use the grass clippings for your composting system.
- Reduce the size of your lawn by placing pavers along the areas where you walk most.
- Plant some trees and shrubs in the middle of the lawn.

If you don't have a clothesline for drying your washed clothes, the lawn is a great place to put one.

Chapter 4

Ten Water-Saving Tips

In This Chapter

- Stopping water waste
- Collecting water before it goes down the drain
- Rating water-saving products
- Recycling greywater
- Gardening water-efficiently

One of the damaging results of climate change is that it makes arid areas even more arid and droughts last much longer. This phenomenon is a huge problem in a dry place like Australia: So much so that in some areas of Australia people are genuinely shocked when water starts falling from the sky. Water is now a precious resource that everyone is beginning to value much more than they used to.

People are slowly but surely accepting the need to change the way they have traditionally used water in their homes and gardens. The fact that you're reading this book means you want some help in getting started, so here are my top-ten tips for saving water and taking pressure off the water supply system.

Turn It Off

If you're like me, you may have some daily routines that result in copious amounts of water being used every day: Showering in the morning or evening (or both), cleaning your teeth, going to the toilet, doing the clothes washing, preparing and cooking your food, rinsing the dishes, and using the dishwasher. All these activities require water.

But the water you waste by leaving the tap or shower running can double your water usage. For every drop of water used, just as much (if not more) goes wasted down the drain simply because people lazily or dreamily leave the tap running. Long showers, too, are a particular problem that many people now focus on cutting down on. So use only what you need and just turn the tap off. Simple as that.

Check the Plumbing

Do you have one of those showers that seems to take an age for the hot water to arrive after you turn the tap on? And do you just let the water run down the drain until you feel some heat and feel comfortable about stepping into the shower? If it takes one minute to warm up, you're simply pouring 5 litres of water down the drain. This major waste of water doesn't have to be so — this problem can usually be fixed by the plumber. And while you're at it, ask the plumber to check for leaky taps and toilets — buckets of water are wasted daily as a result of unattended leaks.

Switch to Water-Saving Taps and Shower Heads

Ask your plumber to check your shower heads and taps. You can actually halve the amount of water you use by switching your shower heads and taps to the many highly rated water-saving products now available on the market. Although most new houses and apartments are outfitted with modern water-saving taps and shower heads, checking with your plumber to see how your current taps and shower heads rate is a worthwhile exercise, just to be sure.

Look at the Ratings

Shower heads and taps aren't the only products with a rating for their water-saving ability. Most household appliances that use water are now rated based on their level of water consumption, similar to the Australian Appliance Energy

Rating Scheme (ERS) that rates appliances for electrical energy consumption. For more about ERS, check out another of my books, *Energy-Saving Tips For Dummies* (Wiley Publishing Australia Pty Ltd).

The Water Efficiency Labelling and Standards Scheme (WELS) rates washing machines, dishwashers, toilets, showers and taps for their water consumption — measured in litres of water used each wash for machines, each minute for showers and taps, and each flush or half flush for toilets. So you can now buy appliances and bathroom accessories knowing the comparative water use qualities of different products. Chapter 2 gives you more information about WELS.

Catch It Before It Goes Down the Drain

One way of overcoming some of the problems of water waste is to capture the water you're using or missing before it goes down the drain. You can then plan to use this water for many daily routines that don't require fresh water from the tap such as watering the pot plants or garden, or rinsing the dishes before they go into the dishwasher. For instance, take a bucket with you to the shower or place a bowl in the sink to capture the water you don't use while waiting for the water to heat up. For more water-saving tips in the bathroom and kitchen, refer to Chapter 2.

Fixing Your Toilet Habits

A three-star WELS rated toilet uses 6 litres of water every flush — the equivalent of filling an average-sized bucket. And while I'm not sure how often you visit the toilet every day, working out how much water you use to cater to your daily toiletry needs isn't difficult. Simply schooling yourself and your family members to press the half-flush button rather than the full-flush button halves this amount instantly! I know some people who even use the water they collect from their showers and taps to use flushing the toilet rather than pressing either of the flush buttons. Chapter 2 covers more water-saving ways in the bathroom.

Save Every Last Drop

In the old days, many houses had water tanks to capture the rain because they weren't connected to a water mains system. The tank was their main source of water for household use. Water tanks are again becoming popular as a great alternative to relying solely on water from the tap. Modern water tanks come in all shapes and sizes to work in with any available space you have. They're especially popular as an alternative water source for the garden.

Recycle Your Greywater

An increasing number of households with gardens and backyards are deciding to go a little bit further than collecting surplus water in buckets. They're installing greywater diversion or treatment systems, which divert hundreds of litres of water otherwise sent down the drain from baths, sinks, laundry tubs, washing machines, dishwashers and kitchen sinks into the garden. These systems are becoming very popular in an era where water restrictions focus on reducing the amount of garden watering in most cities and towns. Check out Chapter 1 for all about greywater treatment systems.

Make Better Use of the Lawn

The amount of space dedicated to grass lawns in suburban Australia is mind boggling. And even more mind boggling is the amount of water some home owners use to maintain pristine looking lawns that add no real value to the local ecosystem. It is little wonder that the government's main approach in restricting household water use has been to force people to turn their sprinklers off during certain times of the year. You can greatly reduce your water use by converting some of your lawn into a fruit and vegetable garden or by planting native perennials that don't require nearly as much water as a lawn. Chapter 3 shows you how.

Garden Naturally

With just a bit of research and planning, you can set up a garden that reduces the need for constant watering. The lowest maintenance garden always has plenty of food plants that are native to your area. Native fruit trees are particularly good because they're perennial — they keep producing year in, year out with minimal attention because they're pest resistant and work well if your garden is in a dry climate. Chapter 3 discusses the varying types of natural gardening techniques.

Index

A

acacias, 49
active water-collection heaters, 31, 32
agricultural water shortages, coping with, 22–4
agriculture, efficient water use, 23
algae blooms, 9
Australia, adapting to a drier climate, 20–2
Australian Greenhouse Office, 32
Australian Water Supply Resources, 14

B

banksias, 49
batch composting toilets, 27
bathrooms, 25–8
  don't let taps run, 26
  reducing showering time, 26
  reducing water usage, 13, 26–7
  water-saving shower heads and taps, 26
bio-intensive farming, 43
biodynamic gardening, 42
black water, 16
boosters (for solar water heaters), 32
boronias, 49
bottlebrushes, 49
bottled water, 19–20
  reasons for purchase, 19
  unsustainability, 20

C

calcium in the soil, 40
callistemons, 49
chickens, keeping, 44
chlorine, 8
climatic conditions, 36, 37
closed circuit system (passive water-collection heaters), 30
clothesline, 51
cold composting, 46–8
compost, making your own, 45–6
compost bins, 45
compost heaps, 36
composting, 43. *See also* worm farms
  cold, 46–8
  harvesting the compost, 46
  ingredients, 45
  maintaining the compost, 45
  warm, 46
composting toilets, 27
continuous composting toilets, 27
crops, water use, 23
cryptosporidium, 18

D

dams, 8, 9, 20
desalination plants, 22
detergents, washing, 29
direct greywater diversion system, 17, 36
dishwashers, 28–9
domestic greywater treatment systems, 17–18
drinking water
  bottled water, 19–20
  rainwater tank water, 18–19
  tap water, 18–19
drought-tolerant species, 23
dyandras, 50

E

eco-friendly washing detergents, 29
environmental impact, desalination plants, 22
eucalypts, 50

• F •

farming
  bio-intensive, 43
  Fukuoka, 42
fertilisers, organic, 40
filtration plants, 8
fluoride, 8, 18, 20
front-loading washing machines, 29
fruit trees, 36
Fukuoka farming, 42

• G •

garden design, 34–8
  landscaping, 35–6
  planning what to plant, 36–7
  working with the lay of the land, 37–8
garden health, maintaining, 39–40
gardening
  biodynamic, 42
  bio-intensive farming, 43
  Fukuoka farming, 42
  organic, 42
  permaculture, 42
gardens, 33–51
  composting waste, 43, 45–6
  discouraging pests, 39
  growing food naturally, 42–3
  keeping chooks, 44
  lawns, 13, 50–1
  maintaining healthy soil, 39
  native plants and shrubs, 48–50
  organic fertilisers, 40
  organic insecticides, 40
  planning what to plant, 36–7
  reducing water usage, 13
  reducing weeds, 39
  resources, 41
  space available for, 37
  sustainable, 38–43, 51
  water usage, 37, 38–9
genetically modified (GM) plants, 23
giardia, 18
greenhouse gas emissions, 22, 32
grevilleas, 50
greywater, 16, 39
greywater diversion systems, 17
greywater recycling, 11, 16–18, 29, 39
greywater treatment systems,
    domestic, 17–18
groundwater, 24

• H •

healthy soil, 39, 40–1
herbs, 36
honey myrtles, 50
hot water
  cost of, 32
  solar, 30–2

• I •

insecticides, organic, 40
irrigation, 9, 23
irrigation channels, concreting and
    covering, 23

• K •

kitchens
  dishwashers, 28–9
  reducing water usage, 13, 28–9
  rinsing and preparing food, 28
  washing dishes by hand, 28

• L •

landscaping, 35–6
laundry, 29
  eco-friendly washing detergents, 29
  reducing water usage, 13, 29
  washing less often, 29
  washing machines, 29
lawnmowers, 51
lawns, 13, 50–1
  unsustainability, 51
leaking taps, 12
leaking toilets, 26

• M •

magnesium in the soil, 40
melaleucas, 50
mulching, 39

• N •

native fruits, 36
native plants and shrubs, 48–50
  choosing, 49–50
  sources of, 50
nitrogen in the soil, 40
nutrients, soil, 40–1

• O •

open circuit system (passive water-collection heaters), 30
organic fertilisers, 40
organic gardening, 42
organic mulch, 39
oxygen in the soil, 41

• P •

paperbarks, 50
passive water-collection heaters, 30–1, 32
pathogens in drinking water, 18, 19
perennial plants, 42
permaculture, 42
phosphorus in the soil, 40
planning the garden, 36–7
plants for the garden, 36–7, 48–50
potassium in the soil, 40
poultry keeping, 44
pumped greywater diversion system, 17

• R •

rainfall, 10, 38
rainwater harvesting, 15–16
rainwater tank water, drinking quality, 18–19
rainwater tanks, 9, 11, 15–16, 38–9
  shapes and sizes, 15
recycled water, 21
recycled water systems, 16–18
recycling
  greywater, 11, 16–18, 39
  urban wastewater, 21
remote communities, water supplies, 24
reservoirs, 8
reusing wastewater, 11, 21, 39
river ecosystems, 8
run-off, 10–11
rural areas
  efficient water use, 23
  water crisis, 22–3
rural towns, water delivery to, 9

• S •

salinity, 8
sewage, 10, 11, 21
sewerage, 10
shade, garden, 37
shower heads, water-saving, 26
shower time, shortening, 26
siltation of rivers and streams, 23
slope of your land, 37
soft fruit crops, 36
soil nutrients, 40–1
soil types, 38
solar energy, 22
solar power systems, 32
solar water heaters, 30–2
  boosters, 32
stormwater run-off, 10–11
sulfur in the soil, 40
sustainable gardens, 38–43, 51
swimming pools, 13

• T •

tap water, safety of, 18–19
taps
  aerators for, 12
  don't let them run, 26
  leaking, 12
  water-saving, 26
toilets
  dual-flush, 26
  with integrated hand basin, 27
  leaking, 26
  waterless, 27

• U •

underground water sources, 8
United Nations Environment Program, 21
urban areas, water delivery to, 8
urban wastewater recycling, 21

• V •

vegetables, 36
  growing your own, 42
  organically grown, 42

• W •

warm composting, 46
washing detergents, eco-friendly, 29
washing machines, 29
  using greywater from, 29
washing up, 28
wastewater, 9–10
  components, 10
  recycling from urban areas, 21
  reusing, 11, 21, 39
  sources, 11
water. *See also* water usage
  as a precious commodity, 8–11
  bottled water, 19–20
  tap water safety, 18–19
water-borne diseases, 18, 19
water-collection heaters
  active, 31, 32
  passive, 30–1, 32
water consumption. *See also* water usage
  'carrot and stick approach' to changing behaviour, 12
  government approaches to reducing, 11–12
  reducing, 12–14
water delivery
  to homes and businesses in urban areas, 8
  to rural towns, 9
  unsustainability of current methods, 9
Water Efficiency Labelling and Standards (WELS) Scheme, 27
water filters, 19
water losses, 10
water pipes, 8, 9
water-rating labels, 28
water recycling, 11, 16–18, 21, 39
water restrictions, 12
water re-use, 39. *See also* greywater recycling; recycled water systems
water-saving shower heads and taps, 26
water scarcity, adopting a drier lifestyle, 11–18
water stress, 8
water supplies
  quality, 18, 19
  remote communities, 24
  technological innovation in management, 11
water supply agencies, 14
water tanks. See rainwater tanks
water usage, 10
  bathrooms, 13, 25–8
  gardens, 13, 38–9, 39
  kitchens, 13, 28–9
  laundry, 13, 29
  rural areas, 22–3
water waste, reducing, 13–14
wattles, 48, 49
weed reduction, 39
worm compost material, harvesting, 48
worm farms, 36
  buying farms or boxes, 46–7
  feeding your worms, 47
  setting up, 46–8
  types of worms, 47
worm juice, 47, 48

Notes

Notes

Notes

Business

Australian Editions

1-74031-109-4
$39.95

1-174031-146-5
$39.95

1-74031-004-7
$39.95

1-73140-710-5
$39.95

1-74031-166-3
$39.95

1-74031-041-1
$39.95

0-7314-0541-2
$39.95

0-7314-0715-6
$39.95

Reference

0-7314-0723-7
$39.95

1-74031-157-1
$39.95

0-7314-0721-0
$39.95

Gardening

1-74031-007-1
$39.95

Australian Editions

Technology

1-74031-086-1
$39.95

0-74031-160-4
$39.95

1-7403-1159-0
$39.95

1-74031-123-X
$39.95

Cooking

1-74031-010-1
$39.95

1-74031-008-X
$39.95

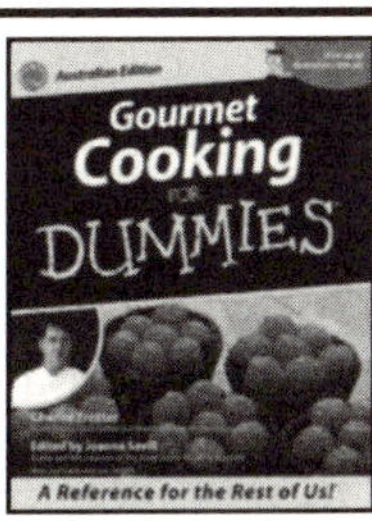

1-74031-040-3
$39.95

Pets

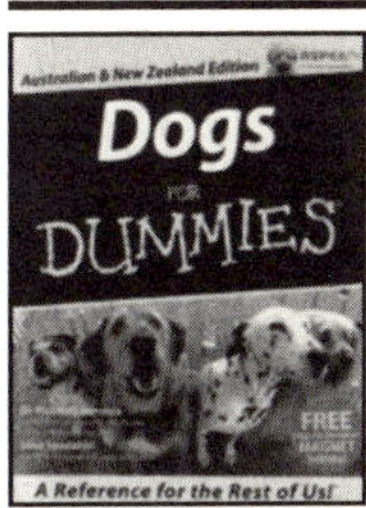

1-74031-028-4
$39.95

Parenting

1-74031-103-5
$39.95

1-74031-042-X
$39.95

Health & Fitness

1-74031-143-4
$39.95

1-74031-140-X
$39.95

Health & Fitness Cont.

Australian Editions

1-74031-122-1
$39.95

1-74031-135-3
$39.95

1-74031-054-3
$39.95

1-74031-009-8
$39.95

1-74031-011-X
$39.95

1-7403-1173-6
$39.95

1-74031-035-7
$39.95

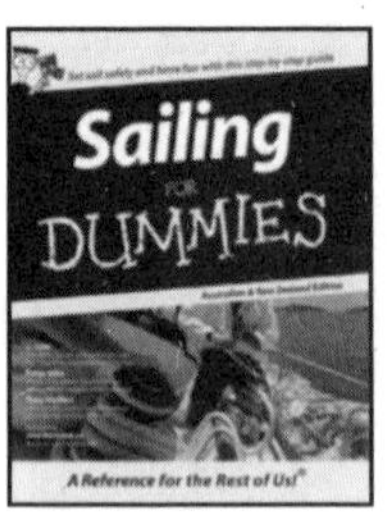

1-74031-146-5
$39.95

1-74031-059-4
$39.95

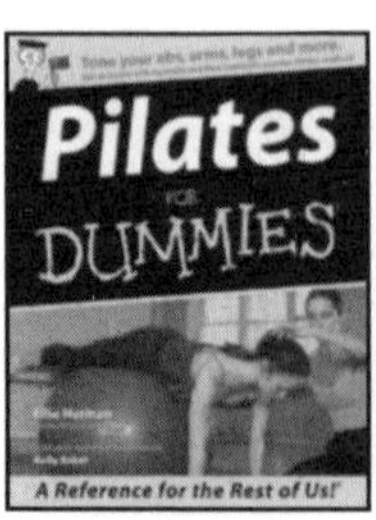

1-74031-074-8
$39.95

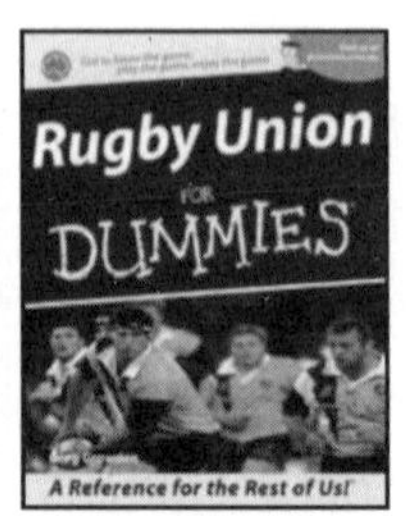

1-74031-073-X
$39.95

1-74031-094-2
$39.95